Emergency Vehicles

Ambulance

Chris Oxlade

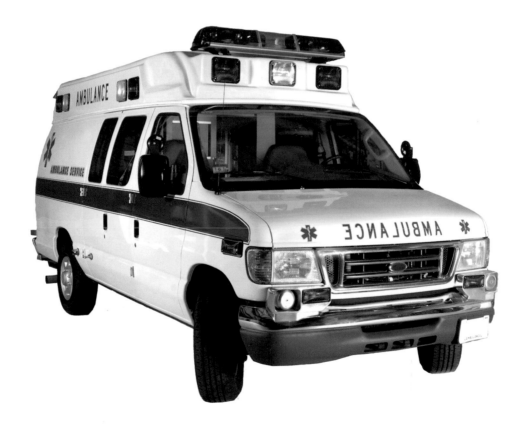

QEB Publishing

This edition published by Scholastic Inc., 557 Broadway,
New York, NY 10012, by arrangement with
QEB Publishing, Inc., 3 Wrigley, Suite A, Irvine, CA 92618

www.qeb-publishing.com

Library of Congress Cataloging-in-Publication Data

Oxlade, Chris.
  Ambulance / Chris Oxlade.
      p. cm. -- (QEB emergency vehicles)
  Includes index.
  ISBN 978-1-59566-975-9 (hardcover)
  1. Ambulances--Juvenile literature. 2. Ambulance service--
Juvenile literature. I. Title.
  TL235.8.O94 2010
  629.222'34--dc22
                              2009005877

Printed and bound in China

ISBN: 978-1-59566-242-2 (paperback)

10 9 8 7 6 5 4 3 2

Author Chris Oxlade
Project Editor Eve Marleau
Designer Susi Martin

Publisher Steve Evans
Creative Director Zeta Davies
Managing Editor Amanda Askew

Picture credits
(t=top, b=bottom, l=left, r=right, c=center, fc=front cover)

Alamy
7r Juice Images Limited; 13r Stephen Bisgrove; 14–15 Trinity
Mirror / Mirrorpix; 15r Stan Kujawa; 16–17 imagebroker; 18l
Topcris
Corbis
18–19 James R. McGury/U.S. Navy/Reuters
Getty
10–11 Taxi
Photolibrary
5r Juice Images
Shutterstock
1 Leonid Smirnov; 4–5 travis manley; 8l Drimi; 9 Pawel Nawrot;
12–13 Cary Kalscheuer; 17r Robert Cumming;
Rex features
12–13 Sipa Press
www.ukemergency.co.uk
6l, 6–7, 20–21, 21r

Words in **bold** can be found in the glossary on page 23.

Contents

What is an Ambulance?

An ambulance is an **emergency** vehicle. It carries sick and injured people to the hospital. Inside an ambulance, there is **medical equipment** to treat patients as the ambulance drives along.

Most ambulances are vans that have been changed to carry people and medical equipment.

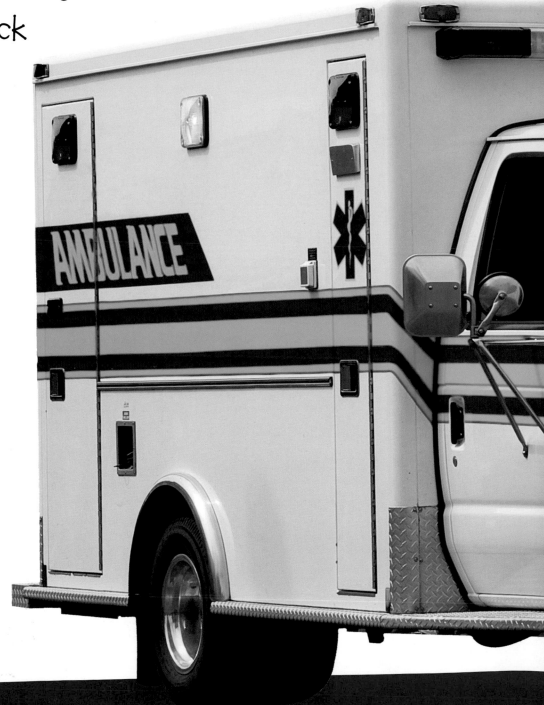

An ambulance has a **crew** on board. They give **first aid** to patients at the scene and inside the ambulance.

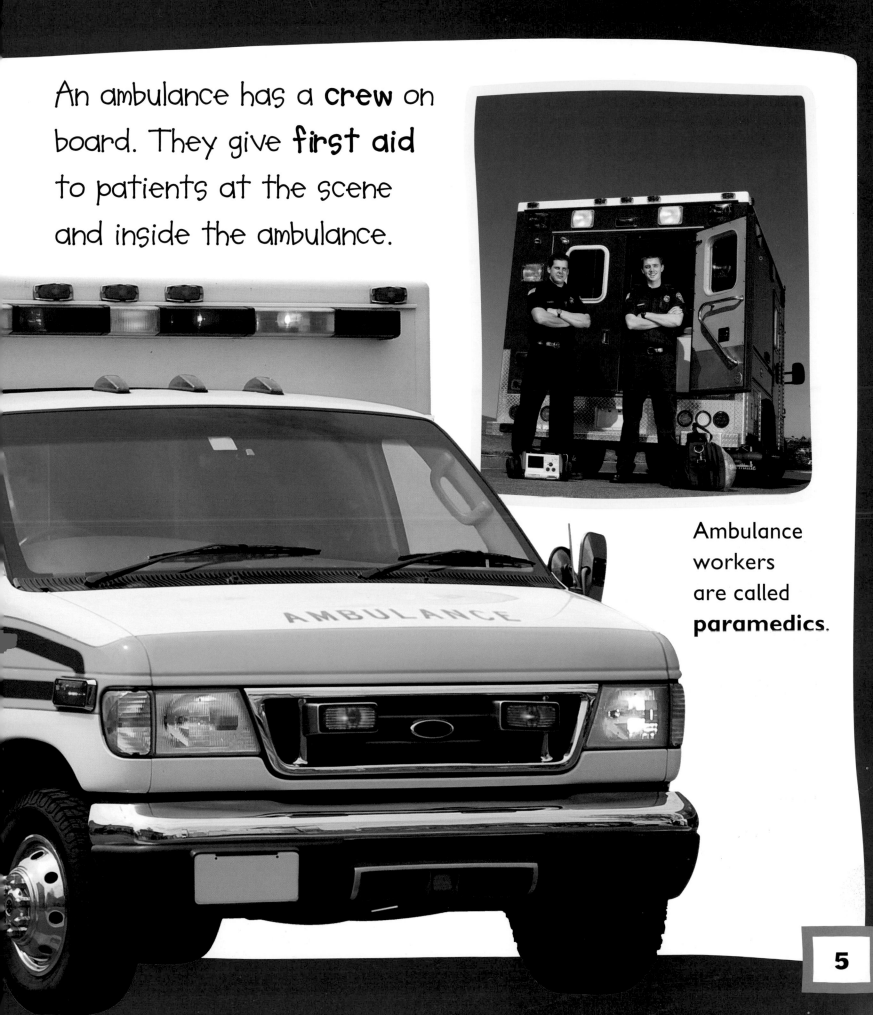

Ambulance workers are called **paramedics**.

Inside the **Cab**

On the way to an emergency, the crew sits in the **cab** of the ambulance. In the cab is a **satellite navigation system** to show the crew the way to an emergency scene.

This shows the inside of an ambulance in the UK. A navigation system guides the crew to an emergency as fast as possible.

The crew has a **two-way radio** to talk to the hospital. Some ambulances also have a computer to send information to the hospital.

Motorcycle ambulances have radios and navigation systems, too.

Ambulance Equipment

The inside of an ambulance is full of medical equipment. There is a stretcher for a patient to lie on. It is used to carry patients from the accident into the ambulance.

There is breathing equipment, a **heart monitor,** and other first-aid equipment. There is also a machine called a defibrillator, which can restart a patient's heart.

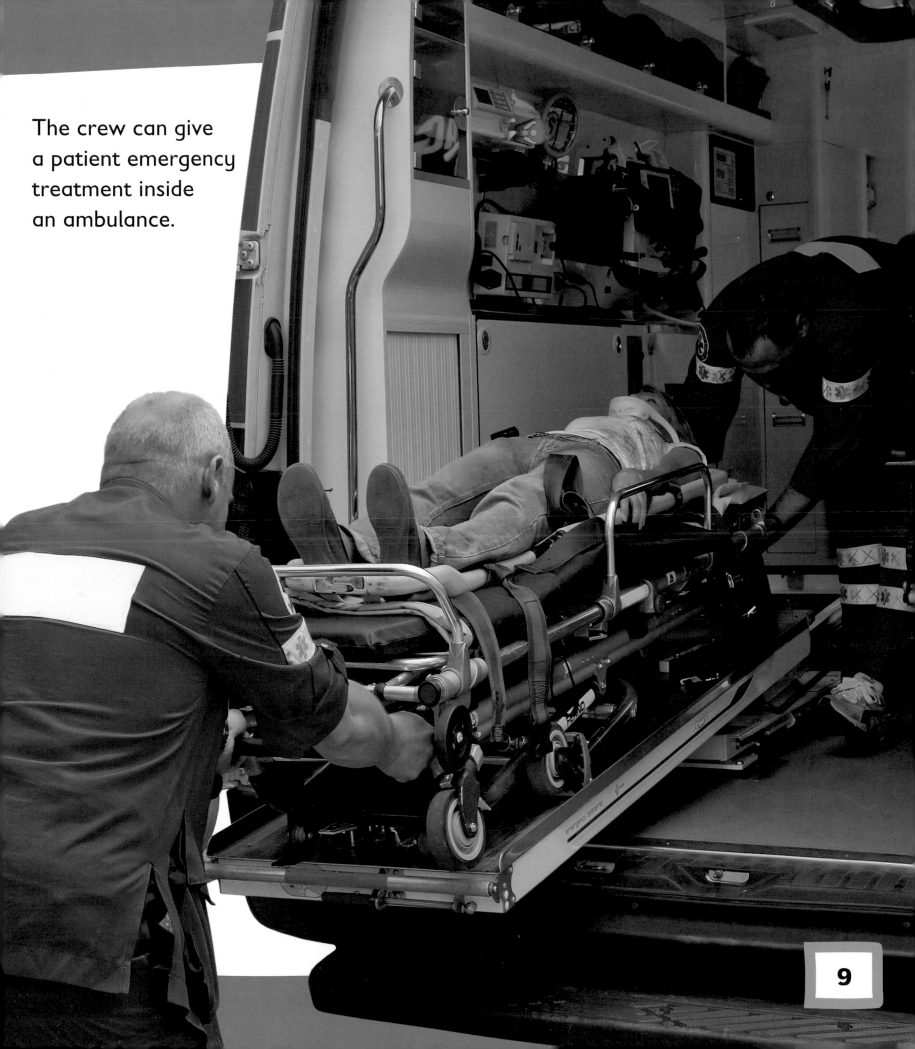

The crew can give a patient emergency treatment inside an ambulance.

9

Lights and Sirens

Ambulances also have flashing lights and noisy **sirens**. They warn other drivers that an ambulance is on an emergency call.

Ambulances often have colorful blocks of paint and **reflective** strips that shine in the dark to make them easy to see.

Sirens make loud screeching noises so that people know an ambulance is coming. Other drivers must move out of the way when they see flashing lights and hear sirens.

Fast Response

Some ambulances are known as rapid-response vehicles. They can reach the scene of an emergency very quickly. Rapid-response vehicles speed through busy **traffic** faster than larger ambulances.

In busy city centers, ambulance crews can reach an emergency quicker by bicycle.

Rapid-response ambulances cannot carry patients to the hospital, but their crews can treat patients with medical equipment. They look after the patients until a regular road ambulance arrives.

W693 MRA

AMBULANCE

Air Ambulances

An air ambulance is called if a patient needs to be taken to the hospital very quickly. In most places, air ambulances are helicopters. They often land on the roof of a hospital to drop off patients.

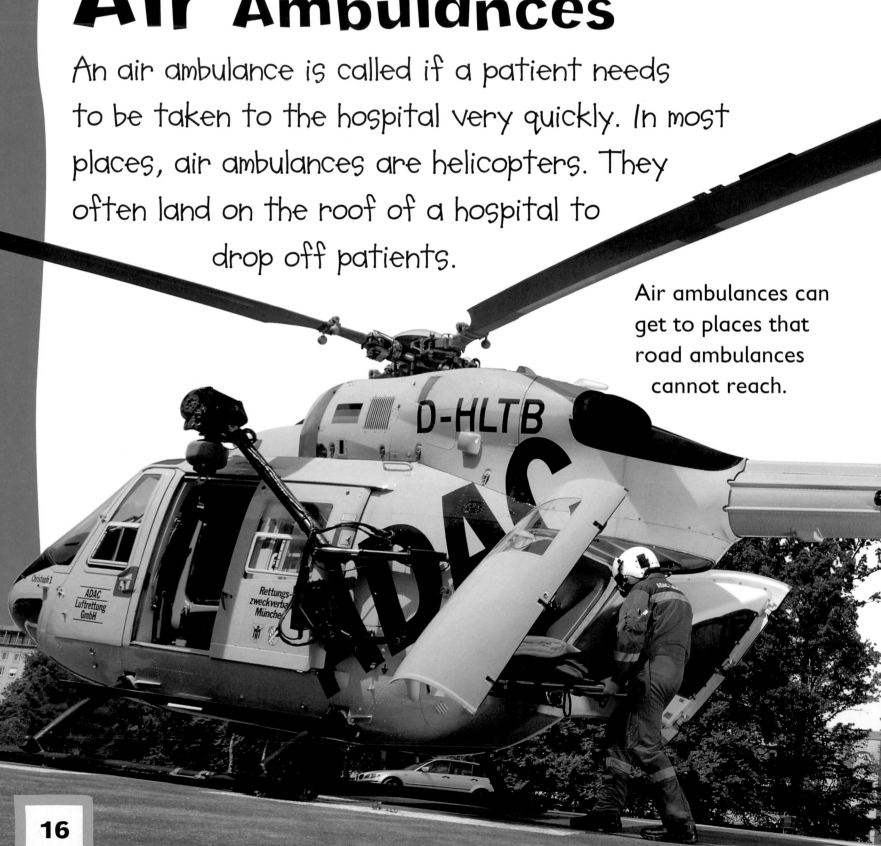

Air ambulances can get to places that road ambulances cannot reach.

Sometimes, air ambulances are **fixed-wing aircraft**. These planes fly faster and farther than helicopters.

Ambulance Ships and Boats

A hospital ship is a ship with a hospital inside. Hospital ships are used mostly by armies to carry injured soldiers.

Hospital ships are normally used near war zones. They always have a red cross on the side.

In places where many people live next to a river, lake, or **waterway**, patients are sometimes carried by ambulance boats.

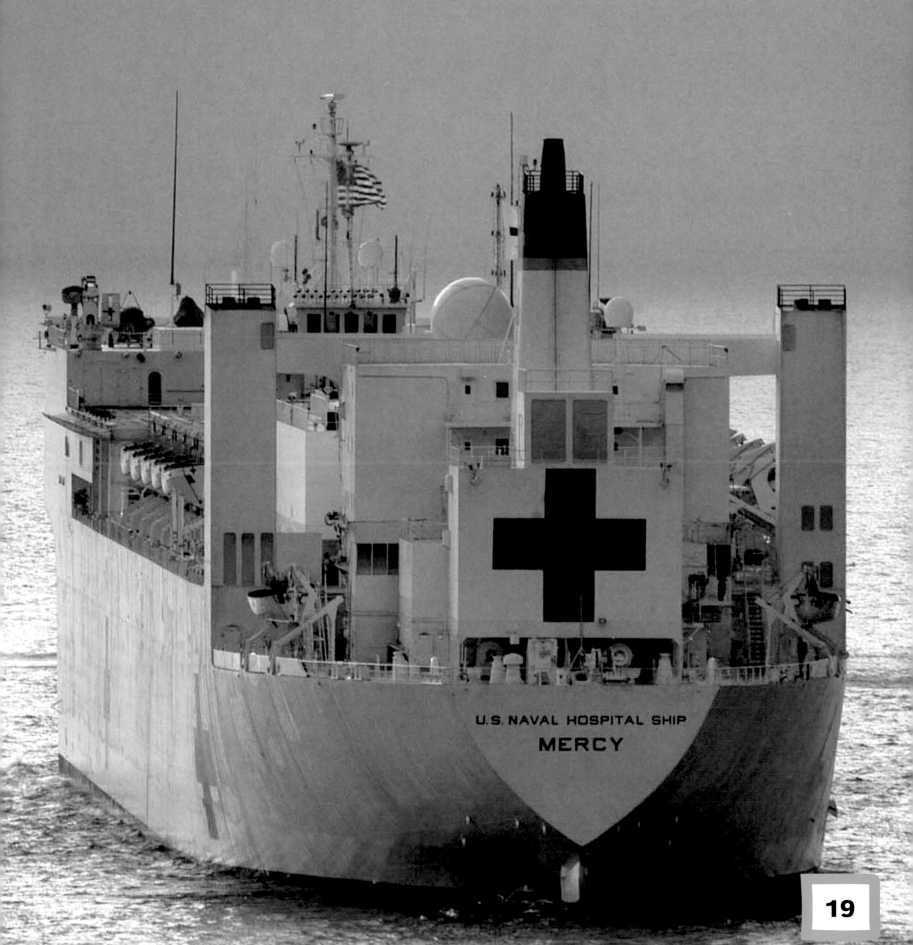

U.S. NAVAL HOSPITAL SHIP
MERCY

Animal Ambulances

An animal ambulance carries sick or injured
animals. It takes them to a vet or animal hospital.
The ambulance carries medical equipment
to treat the animal.

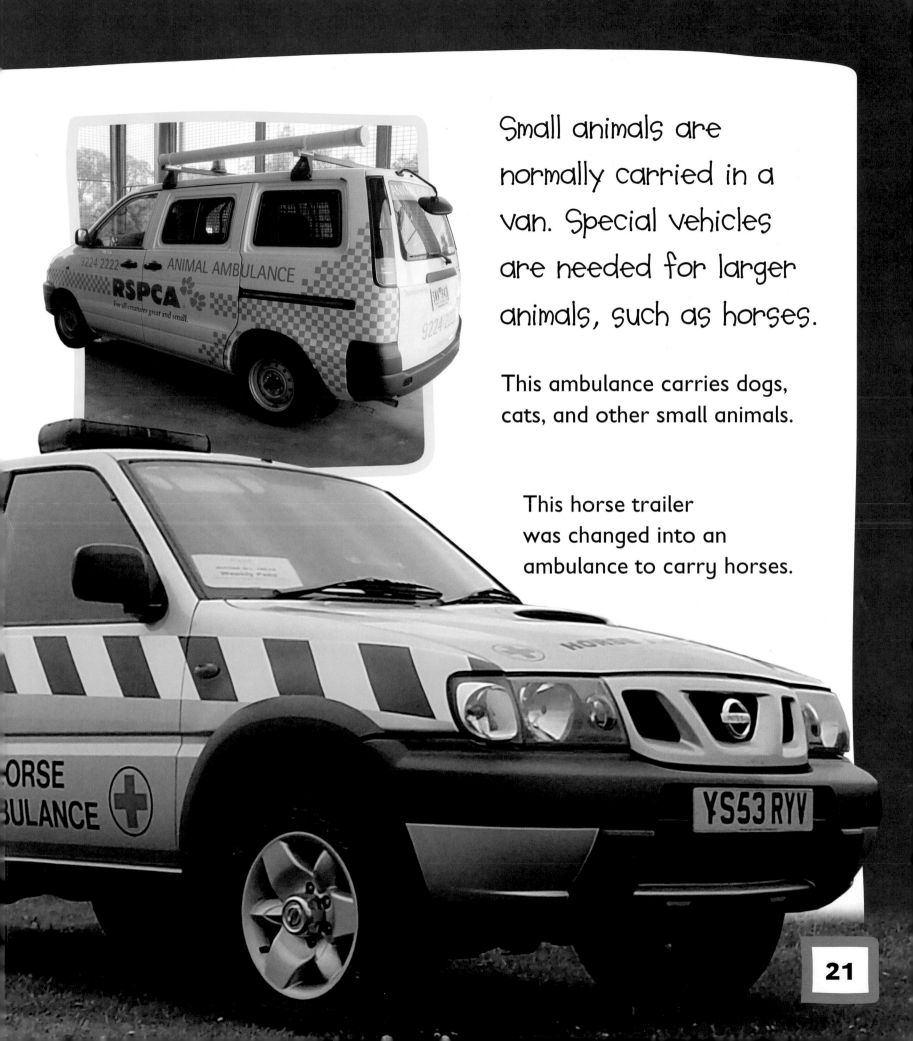

Small animals are normally carried in a van. Special vehicles are needed for larger animals, such as horses.

This ambulance carries dogs, cats, and other small animals.

This horse trailer was changed into an ambulance to carry horses.

Activities

- Which picture shows a air ambulance, an horse ambulance, and a hospital ship?

- Make a drawing of your favorite ambulance. What sort of ambulance is it? Does it have lights and sirens? What color is it?

- Write a story about a medical emergency. It could be anywhere in the world —or even on another planet! What ambulance would you drive? What kind of accident was it? How many people were involved?

- Which of these ambulances would be used to pick up a small animal?

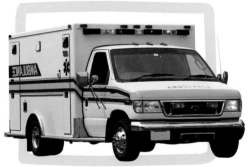

Glossary

Cab
The part of the ambulance where the driver sits.

Crew
A group or team of people.

Emergency
A dangerous situation that must be dealt with right away.

First aid
Medical help given to patients before they are taken to the hospital.

Fixed—wing aircraft
An aircraft with two wings that are fixed to its body.

Heart monitor
A machine that measures your heartbeat.

Medical equipment
Equipment that is used to treat illnesses or injuries.

Paramedic
A person who is trained to give people medical treatment in an ambulance.

Reflective
A material that light bounces off very well.

Satellite navigation
Finding the way from one place to another.

Siren
A machine that makes a loud, screeching noise.

Traffic
Cars, trucks, motorcycles, and other vehicles on the road.

Two—way radio
A radio set that lets people talk to each other.

Waterway
A river or canal that ships travel along.

Index